PROJECT M²: MENTORING YOUNG MATHEMATICIANS

Using Everyday Measures

Measuring with the Meerkats

M. Katherine Gavin, *University of Connecticut*
Tutita M. Casa, *University of Connecticut*
Suzanne H. Chapin, *Boston University*
Linda Jensen Sheffield, *Northern Kentucky University*

Student Mathematician's Journal

Connections to the Common Core State Standards

This material is based upon work supported by the National Science Foundation under Grant No. 0733189 and developed at the University of Connecticut. Any opinions, findings, and conclusions or recommendations expressed in this material are those of the author(s) and do not necessarily reflect the views of the National Science Foundation.

Kendall Hunt
publishing company

www.kendallhunt.com
Send all inquiries to:
4050 Westmark Drive
Dubuque, IA 52004-1840
1-800-542-6657

ISBN 978-1-5249-1219-2

Published in the United States of America

Production Date: 2017
Printed by: LSI
United States of America
Batch number: 441219

17 16 15 14 13

Table of Contents

Table of Contents

As **Speakers** we will:

1. Talk loud enough for others to hear.

2. Turn to talk to the class.

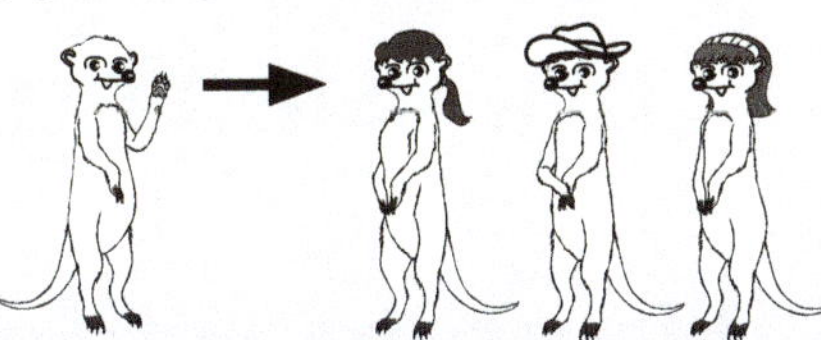

3. Share different ideas.

4. Explain our ideas.

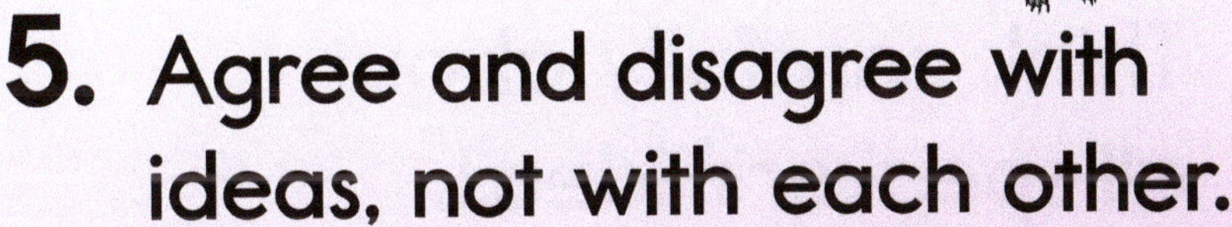

5. Agree and disagree with ideas, not with each other.

As **Listeners** we will:

1. Ask speakers to speak up.
2. Show speakers we are listening.

3. Listen to understand.
4. Ask questions to make sense of the idea.

Did you mean...?

5. Think carefully about all speakers' ideas.

As Writers we will:

Think about the question.

Talk about the answer.

Tell ▶ Tell:

1. all ideas
2. the answer
3. "why"

Thinking Like a Mathematician*

Here is a list of skills mathematicians use every day. See how many you can use in your Student Mathematician's Journal.

1. Work hard to understand and solve problems.
2. Make sense of numbers in problems and use operations with numbers.
3. Defend your math ideas. Explain why you agree or disagree with someone's thinking.
4. Use the math you know to help solve problems in your world.
5. Choose and use the right math tools to help solve problems.
6. Use correct math vocabulary and symbols to explain clearly.
7. Look for and use patterns to help solve problems.
8. Notice if you are using the same math again and again and look for short cuts.
9. Solve a problem in a new way. Ask new questions to investigate.**

*Adapted from the Common Core State Standards: Standards for Mathematical Practice

National Governors Association Center for Best Practices (NGA Center), Council of Chief State School Officers (CCSSO). (2010). *Common core state standards for mathematics.* Washington, DC: Authors. Retrieved from http://www.corestandards.org/the-standards.

** Johnsen, S. K., & Sheffield, L. J. (Eds.). (2013). *Using the common core state standards for mathematics with gifted and advanced learners.* Waco, TX : Prufrock Press.

Student Mathematician: ______________________ Date: __________

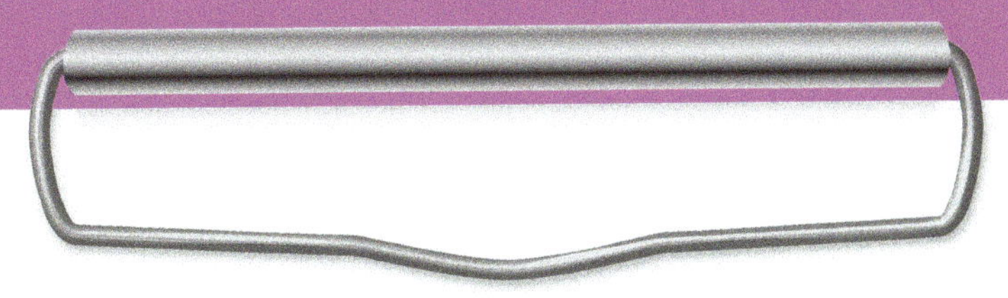

We used this ruler to measure how wide an underground tunnel needs to be for us to fit into it. Dru thinks that this tunnel is 4 inches wide, and Teller thinks it is 5 inches wide. Who is right? Why?

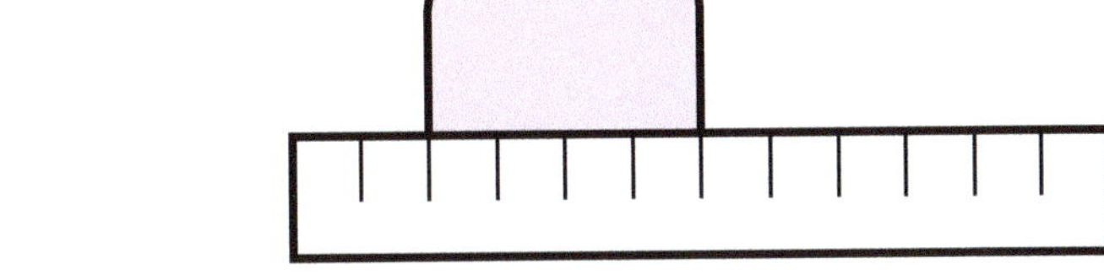

Dear Dru and Teller,

We think that ______________________ is right. The tunnel is ______________________ wide. We know it is ________ inches wide because ____.

The tunnel is not ______________________ wide because ____.

Look what we added to the picture at the top. We hope this helps you build good tunnels. We look forward to hearing from you soon!

______________________,

____

Student Mathematician: Date:

Prickly Measures — Inches

In the Kalahari Desert, we have a lot of cacti. Help us figure out the lengths of these cactus plants. Use your inch ruler to measure to the nearest inch.

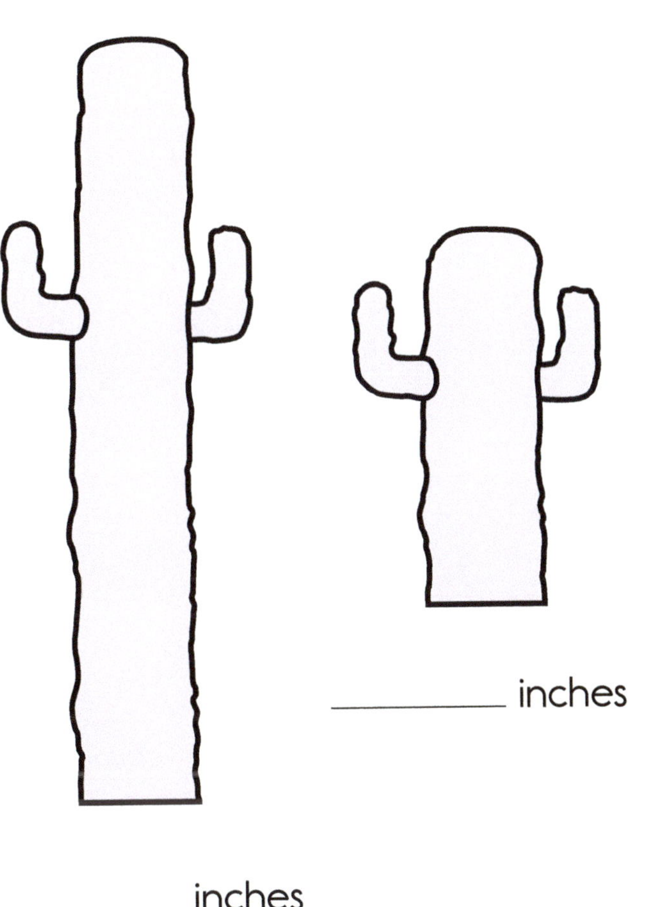

_________ inches

_________ inches

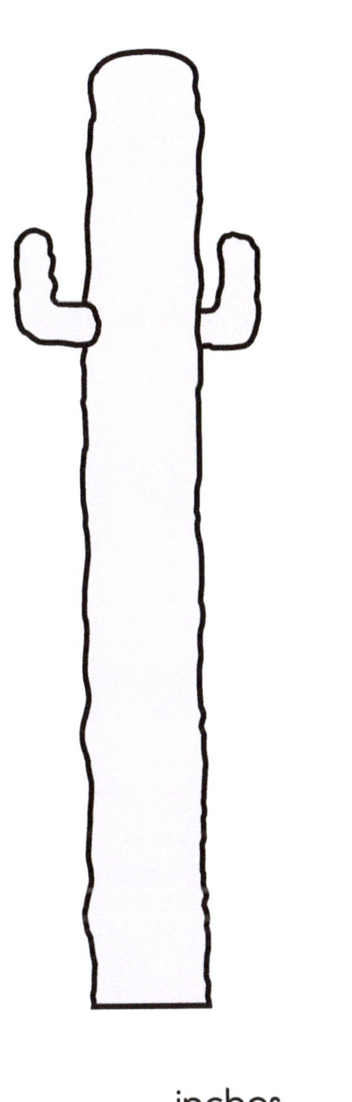

_________ inches

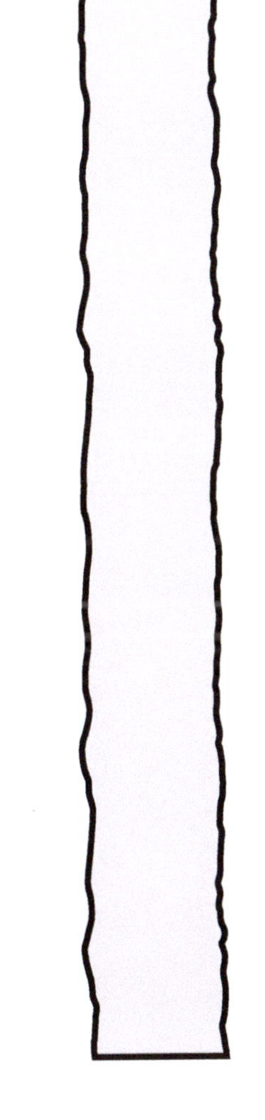

_________ inches

Student Mathematician: ________________________________ Date: ____________

Prickly Measures — Half-Inches

Use your half-inch ruler to measure these cactus plants to the nearest half-inch.

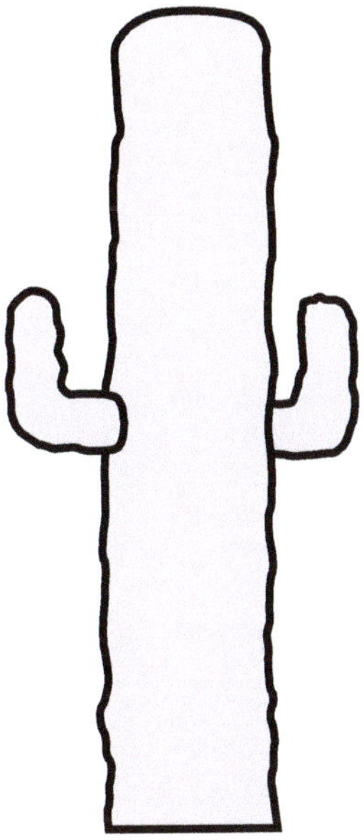

______ half-inches

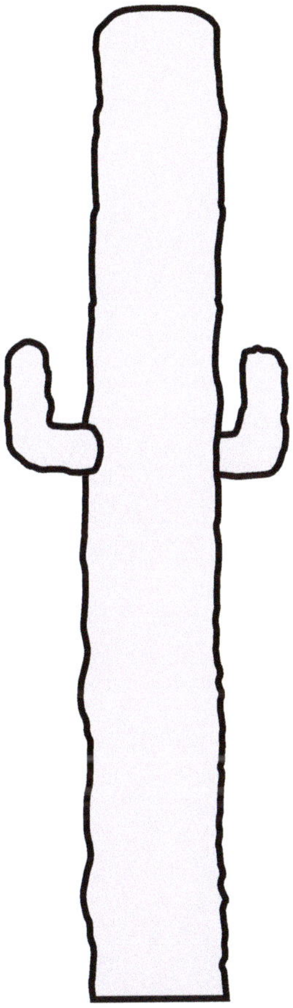

______ half-inches

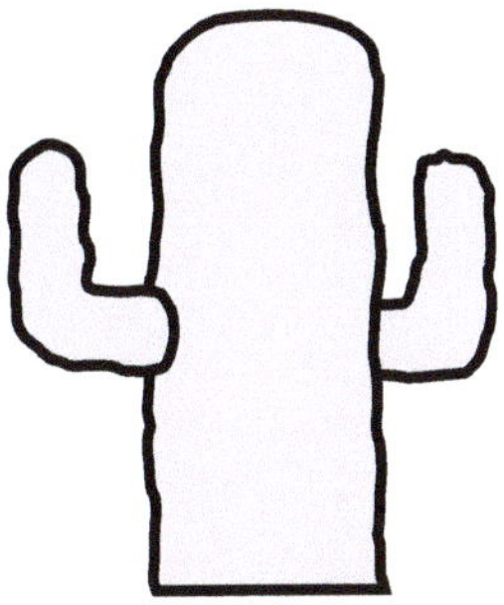

______ half-inches

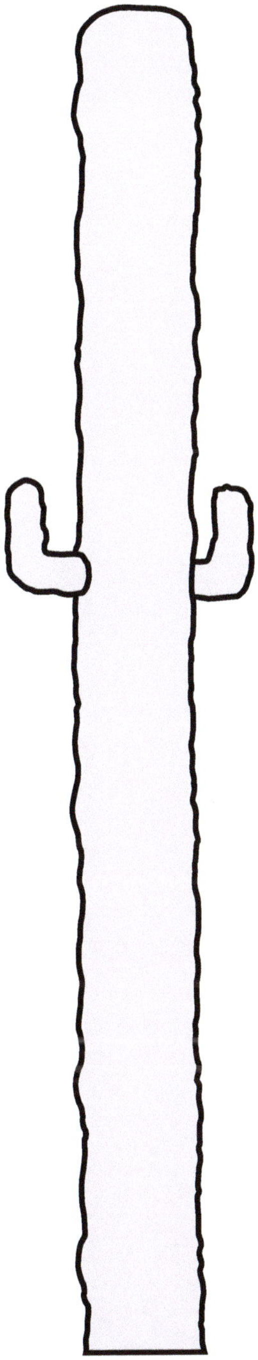

______ half-inches

Student Mathematician: ______________________________ Date: ______________

Getting Ready for the Marble Crash Tests

Help us figure out the lengths of these blades of grass. Use your inch ruler and your half-inch ruler to measure to the closest unit. Write your answers in the spaces below the blades of grass. Remember to label your answers "inches" or "half-inches"!

1. ____________________ inches or ____________________ half-inches

2. ____________________ or ____________________

3. ____________________ or ____________________

4. ____________________ or ____________________

5. ____________________ or ____________________

Student Mathematician: ____________________ Date: __________

Getting Ready for the Marble Crash Tests

Write the lengths of these blades of grass in the spaces below them!

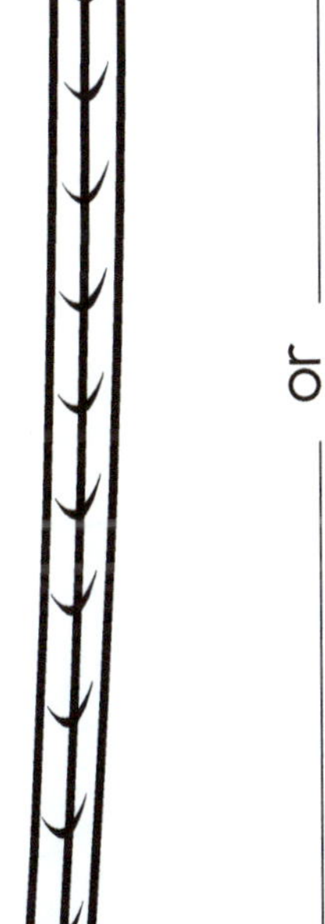

6. ______ or ______

7. ______ or ______

8. ______ or ______

9. ______ or ______

Student Mathematician: ______________________ Date: ______________

Getting Ready for the Marble Crash Tests

Dru used an inch ruler to measure some branches and recorded the measures below. Line up your measurers like the picture. If you used the half-inch measurer, what would the length be in half-inches?

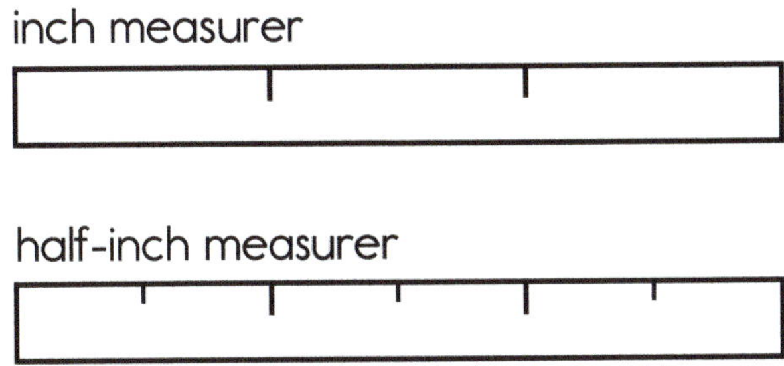

10. 3 inches or ____________ half inches

11. 5 inches or ____________ half inches

12. 2 inches or ____________ half inches

Teller measured some branches and recorded the measures in half-inches. If you used the inch measurer, what would the lengths be in inches?

13. 8 half-inches or ____________ inches

14. 6 half-inches or ____________ inches

15. 2 half-inches or ____________ inches

Student Mathematician: Date:

Marble Crash Tests

Set up your crash test as shown in the diagram below. Both Marble ONE and Marble TWO should be placed in the middle of your rulers (at the 6-inch mark).

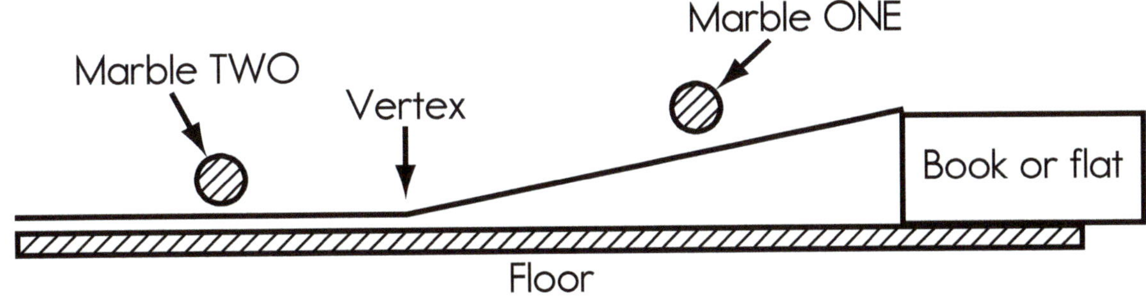

1. When your teacher says to, you will *release* Marble ONE.
2. *Mark* where Marble TWO stops with a sticky note labeled "1".
3. *Measure* the distance from where Marble TWO started to the sticky note using the **inch measurer** and *record* the distance on the chart below.
4. *Measure* the distance from where Marble TWO started to the sticky note using the **half-inch measurer** and *record* the distance on the chart below.
5. *Leave* the sticky note and prepare for Trial 2.

Trial	Distance Traveled	
	Inch Measurer (to the nearest inch)	Half-Inch Measurer (to the nearest half-inch)
1 (height = 1 level)	inches	half-inches
2 (height = 2 levels)		
3 (height = 3 levels)		
4 (height = 4 levels)		
5 (height = 5 levels)		

Student Mathematician: ______________________ Date: ____________

THINK DEEPLY

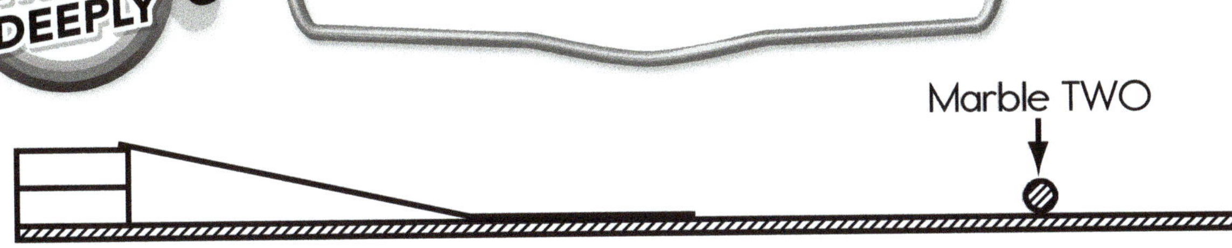

After the marble crash test, Dru measured and said the distance was 14 half-inches.

Teller measured and said the distance was 7 inches.

Who is right? ____________

Why are Dru's and Teller's measurements different?

Student Mathematician: Date:

Centimeter Seatbelts

Help us measure these seatbelts to the nearest centimeter.

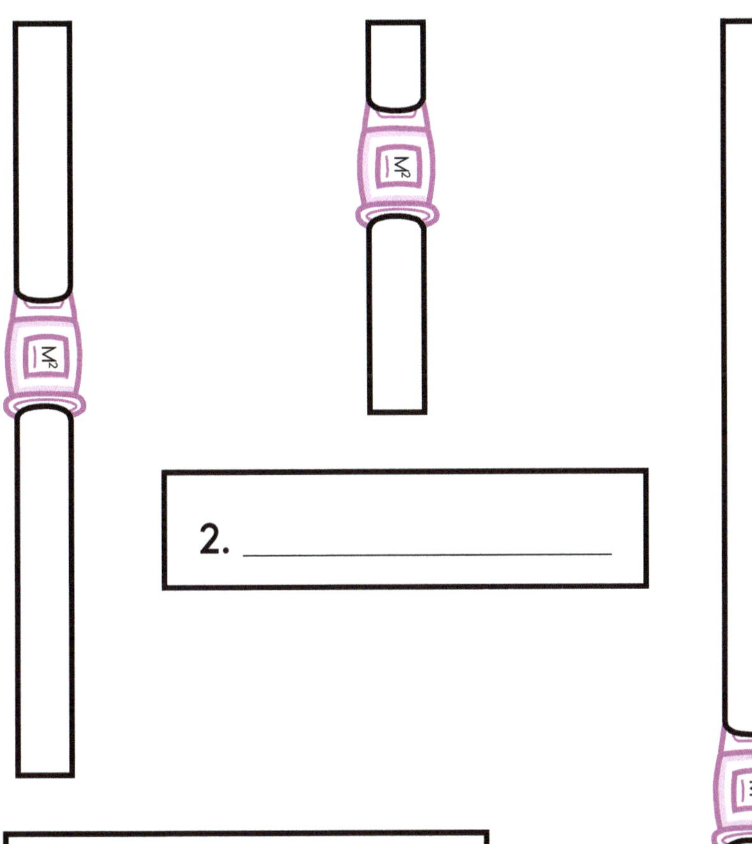

2. ____________________

1. __________ centimeters

4. ____________________

3. ____________________

Student Mathematician: _______________ Date: _______________

Centimeter Seatbelts

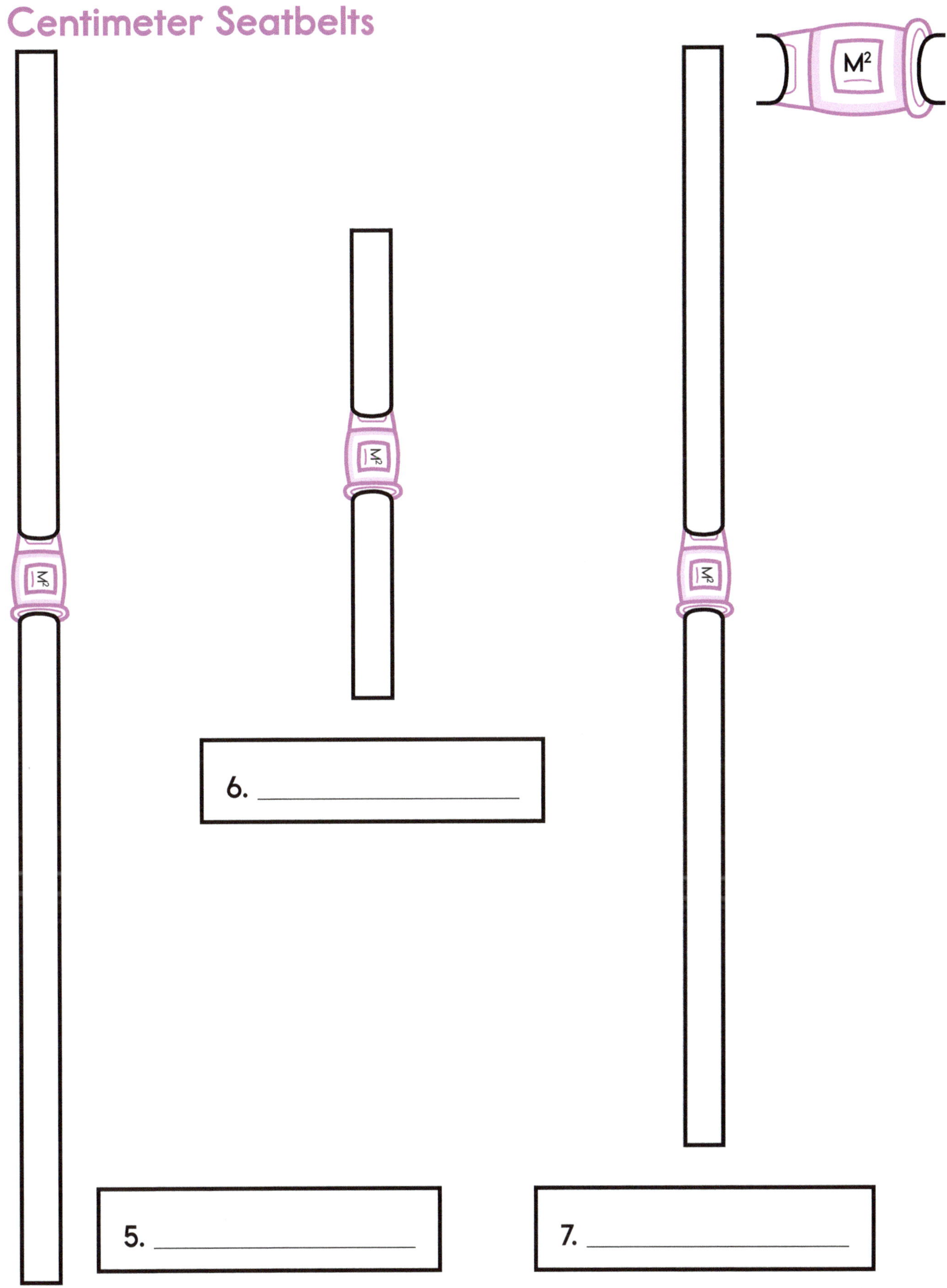

6. ____________________

5. ____________________

7. ____________________

Student Mathematician: Date:

What's the Best Tool?

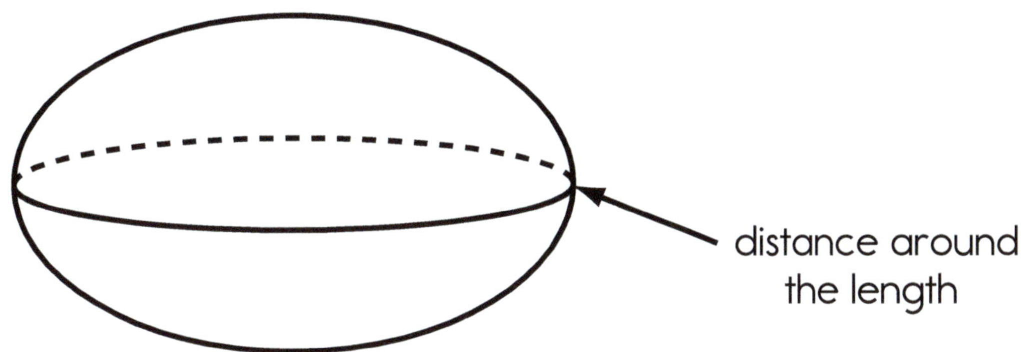

Measurement in centimeters	First Guess	Measure with Cube	Second Guess	Measure with Ruler	Third Guess	Measure with Pipe Cleaner	Fourth Guess	Measure with Tape Measure
Distance around the *length* of the egg								

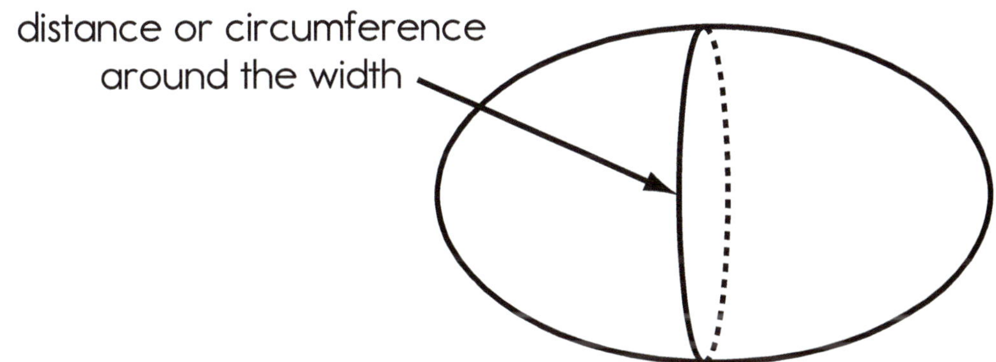

Measurement in centimeters	First Guess	Measure with Cube	Second Guess	Measure with Ruler	Third Guess	Measure with Pipe Cleaner	Fourth Guess	Measure with Tape Measure
Distance around the *width* of the egg								

Student Mathematician: ______________________ Date: __________

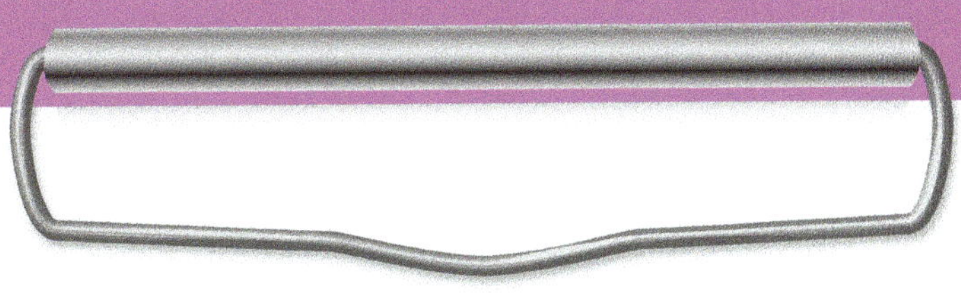

THINK DEEPLY

Dru and Teller need your help with measuring!

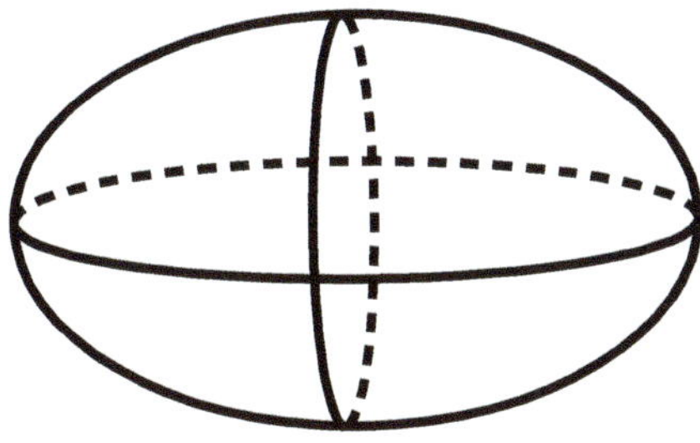

a. Which measuring tool was the best one to measure the circumference of the eggs?

b. Why?

Seatbelt Designers: ______________________________

Quality Control Checkers: ______________________________

Building the Car Seats

Partner Check	Requirements	Quality Control Check
☐	Seatbelt Designers and Quality Control checked to make sure no more than 50 cm of materials were used.	☐
☐	Egg filled with Play-Doh®.	☐
☐ ☐ ☐	The car seat has: a. a part of an egg carton b. the egg carton placed on a piece of foam c. at least two holes punched in the car seat materials to attach the seatbelt	☐ ☐ ☐
☐	The seatbelt is attached to the car seat.	☐

On the back, draw a picture of your seatbelt design. Label all measurements of all parts of the seatbelt.

Student Mathematician: Date:

Measuring Car Seat Materials

We do not have a lot of materials in the desert, so we have to use what is available. Your teacher gave you some materials to use to create a seatbelt.

1. Guess the measurements below in centimeters.
2. Measure your materials in centimeters.
3. Find the difference between your guess and the measurements!

Material	Guess	Measurement	Difference
Foam width	cm	cm	cm
Foam length			
Length of pipe cleaners			
Length of straws			
Length of yarn			

Seatbelt Designers: ______________________________

Quality Control Checkers: ______________________________

How Much Material Did We Use?

Material	No More Than	Designers' Measurements	Quality Control Measurements
Pipe Cleaners	20 centimeters		
Yarn	25 centimeters		
Straws	15 centimeters		
Tape	10 centimeters		
Total	50 centimeters		

Student Mathematician: ______________________ Date: __________

THINK DEEPLY

Dru and Teller want to review your seatbelt design.

a. Tell them what you would change about your design. Why?

b. If Dru and Teller have to use exactly 40 centimeters of materials, list the materials and amounts you would suggest they use. Draw the new seatbelt.

Student Mathematician: _______________ Date: _______________

Sleeping Bag Estimates

1. We estimate that the sleeping bag will cover between _______ square feet and _______ square feet.

   We made two models.

   Our rectangle was _______ feet by _______ feet.

   It was too big / too small (circle one).

   The other rectangle was _______ feet by _______ feet.

   It was too big / too small (circle one).

2. After making models, a better estimate is that the sleeping bag will cover between _______ square feet and _______ square feet.

   We made two models.

   Our rectangle was _______ feet by _______ feet.

   It was too big / too small (circle one).

   The other rectangle was _______ feet by _______ feet.

   It was too big / too small (circle one).

3. On the back, make a drawing of how we decided the number of square feet for our sleeping bag.

Student Mathematician: Date:

Sleeping Bag Areas

1. Which sleeping bag are you measuring?

   ____

2. We predict that the area of the sleeping bag is

   ________ square feet.

3. The area of the sleeping bag is actually ________ square feet.

If you have time, measure a second sleeping bag.

4. Which sleeping bag are you measuring?

   ____

5. We predict that the area of the sleeping bag is ________ square feet.

6. The area of the sleeping bag is actually ________ square feet.

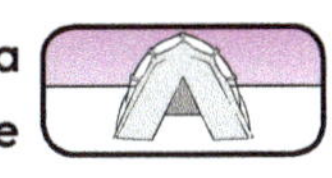

Student Mathematician: ______________________ Date: __________

So Big

Predict first and then measure the area of each object in square feet. Use your paper model of a square foot.

Rectangular Object	Area Prediction (in square feet)	Actual Area (in square feet)	Difference between prediction and actual area
Teacher's Desk			
Student's Desk			

Tent A

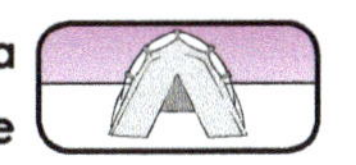

Tent B

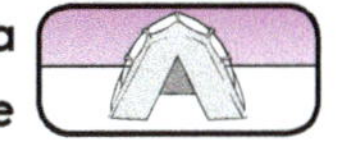

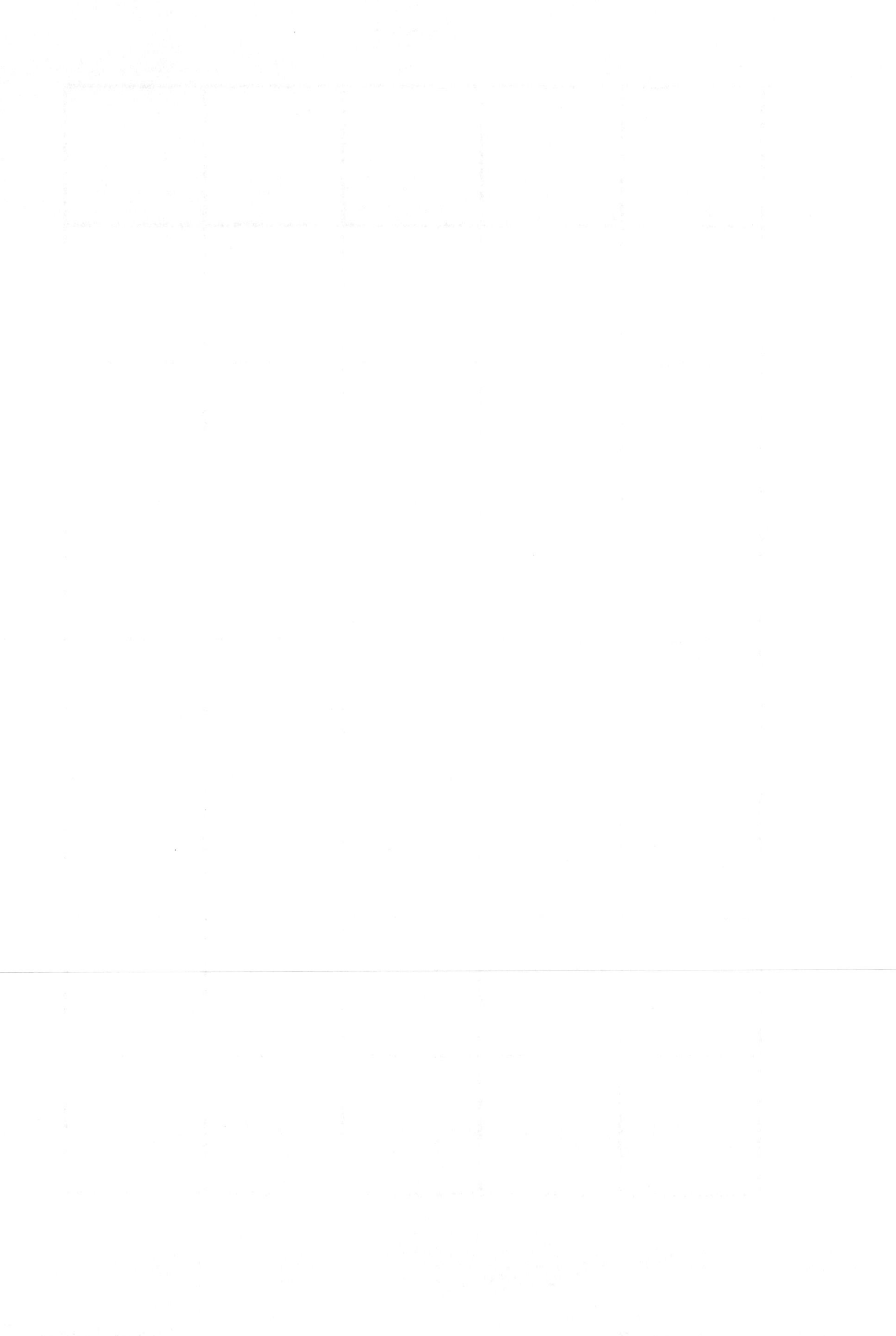

Student Mathematician: ______________________ Date: __________

THINK DEEPLY

Dru and Teller found Tent C. It is 10 feet long and 8 feet wide.

What is the area of Tent C? __________

Dru said that four adults would fit in the tent because each adult needs 18 square feet of floor space (18 + 18 + 18 + 18 = 72). So, there is room to spare. Teller said that he tried and could not get four adults to fit in Tent C.

a. Who is right? __________

b. Explain how you know.

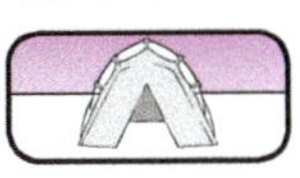

Tent C

Inch Grid Paper

Student Mathematician: ______________________ Date: __________

Build It, Measure It

Use your inch tiles to build these rectangles. Then find the area.

1. 4 inches-by-2 inches

   Area = ______________ *square inches*

2. 3 inches-by-7 inches

   Area = ______________________

3. 6 inches-by-2 inches

   Area = ______________________

4. 4 inches-by-3 inches

   Area = ______________________

5. What do you notice about the rectangles that are 6 inches-by-2 inches and 4 inches-by-3 inches?

   ____

   ____

   ____

Student Mathematician: ____________________ Date: __________

Putting Things in Order

Compare the areas of the "Shapes for Putting Things in Order."

1. **Predict** the order of the shapes from smallest area to largest area on the lines below.

____________	____________	____________
Smallest Area		Greatest Area

2. Cut out the rectangles to **compare** the area of the shapes. Record what you find on the lines below.

____________	____________	____________
Smallest Area		Greatest Area

3. How did you figure out which one had the smallest and which one had the largest area?

4. Measurements with inch tiles:

   A = ________ square inches

   B = ________ square inches

   C = ________ square inches

5. Measurements with inch grid:

   A = ________ square inches

   B = ________ square inches

   C = ________ square inches

Student Mathematician: Date:

Shapes for Putting Things in Order

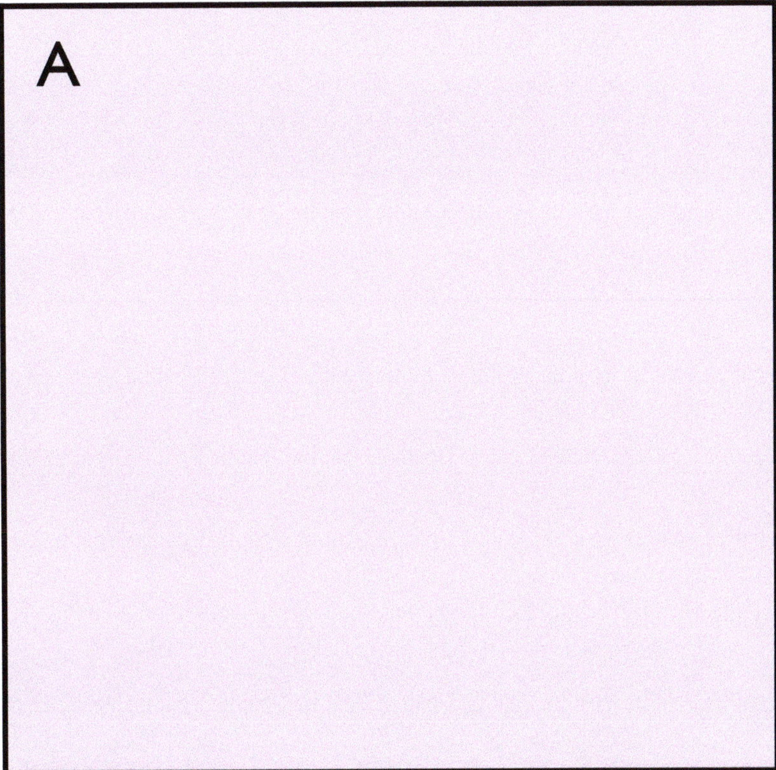

B

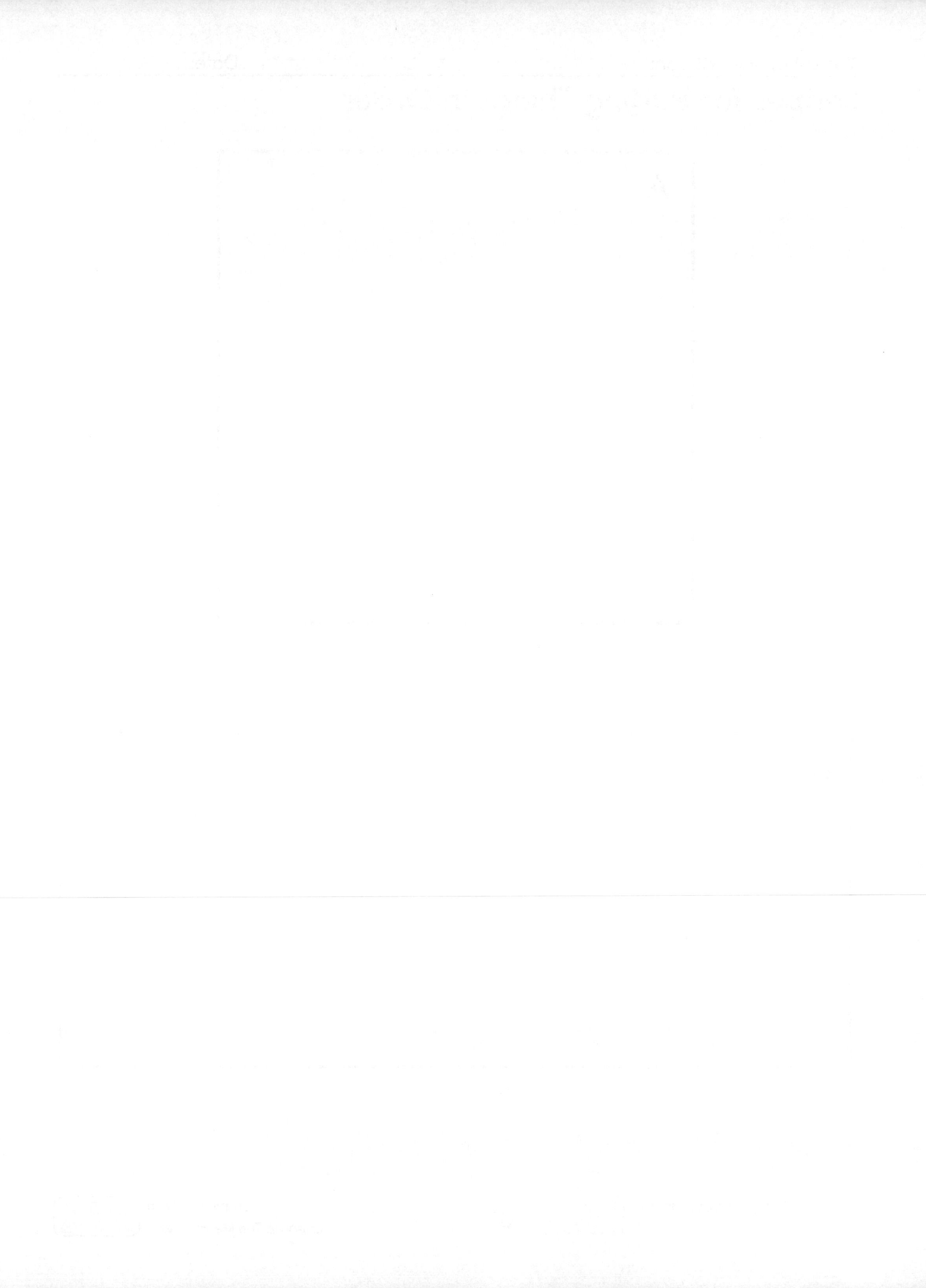

Student Mathematician: Date:

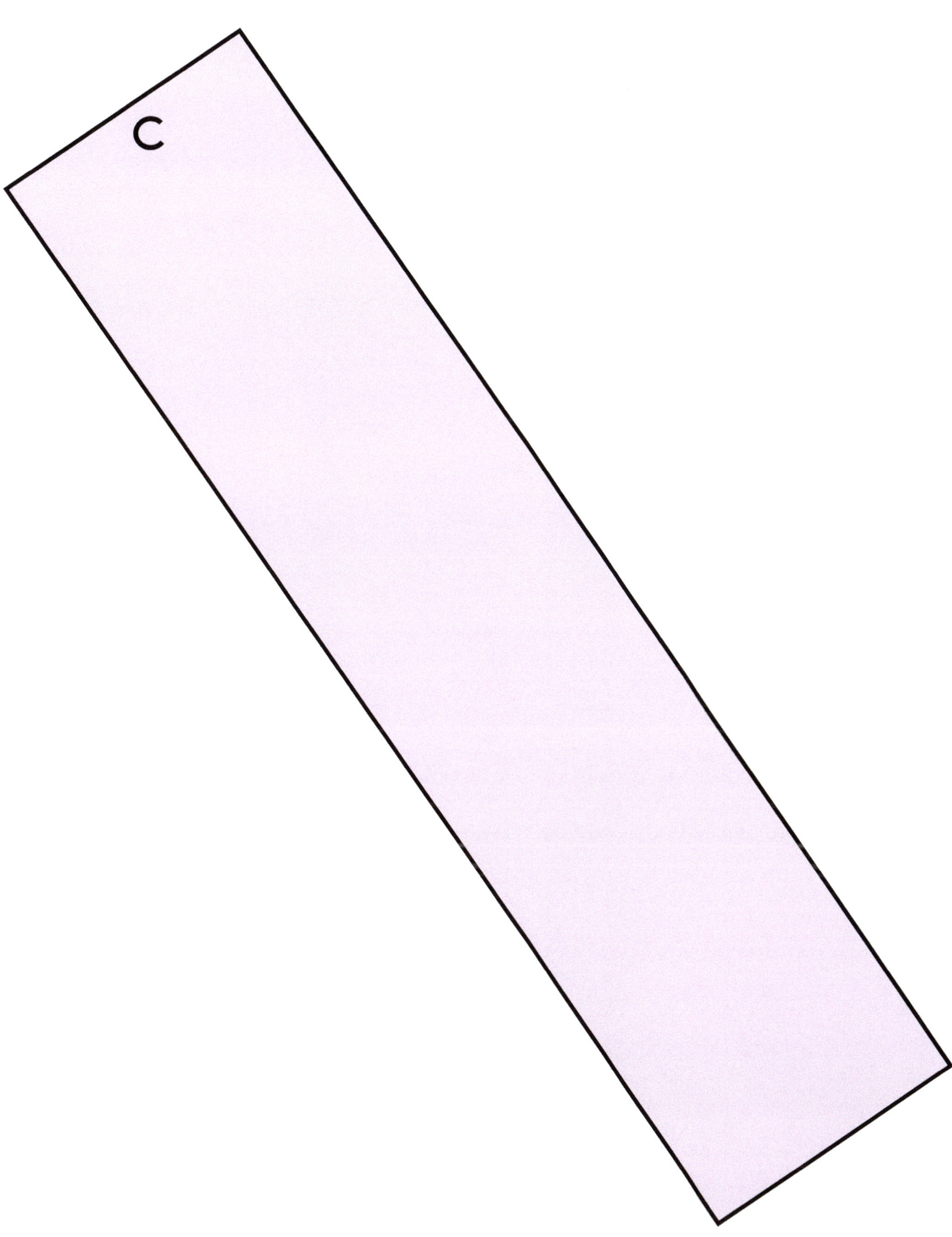

Student Mathematician: ____________ Date: ____________

How Big? — Square Inches

A meerkat has dug these spaces. Predict the area of each and then measure to check. Include the unit.

Shape	Predicted Area	Actual Area	Difference Between Prediction and Actual Measurement
1. A			
2. B			
3. C			
4. D			

Order the shapes from the smallest to the largest areas. ____ ____ ____ ____

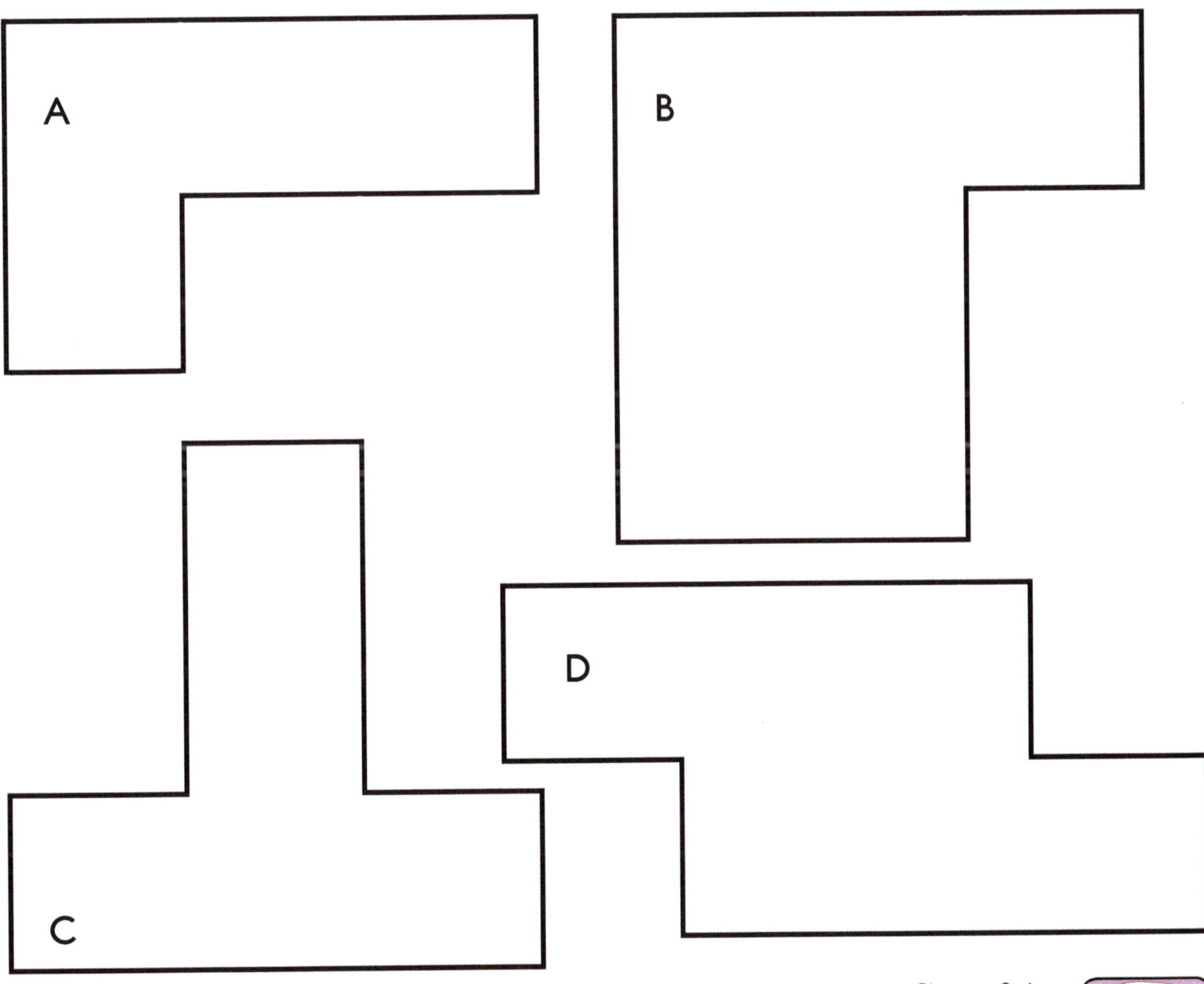

Student Mathematician: ______________________ Date: ______________

How Big? — Square Centimeters

A meerkat has dug these spaces. Predict the area of each and then measure to check. Include the unit.

Shape	Predicted Area	Actual Area	Difference Between Prediction and Actual Measurement
1. A			
2. B			
3. C			
4. D			

Order the shapes from the smallest to the largest areas. ____ ____ ____ ____

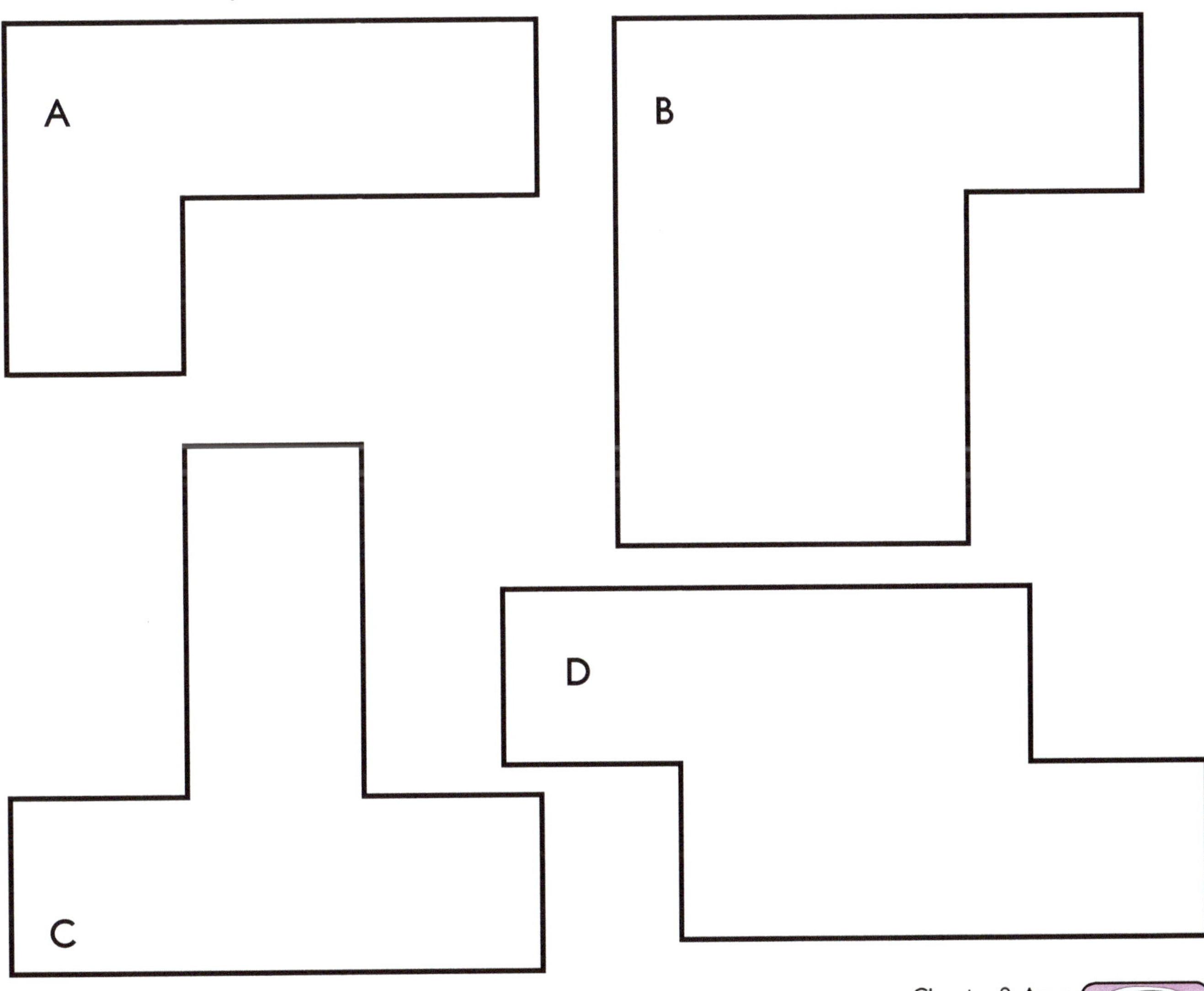

Student Mathematician: Date:

The Scavenger Hunt

**Find things in your classroom or at home to complete the chart.
Use your clear grid paper or paper models to check.**

Area	Object
1. about 12 square inches	
2. about _____ square inches	front cover of my notebook
3. less than 6 square inches	
4. about 20 square inches	
5. about 8 square feet	
6. about _____ square inches	dollar bill
7. about 16 square centimeters	
8. less than 1 square foot	
9. about _____ square centimeters	library card or credit card
10. about 100 square centimeters	

Student Mathematician: ______________________ Date: __________

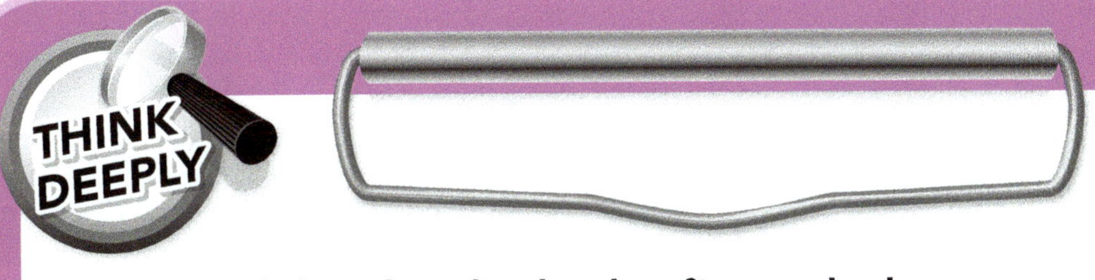

Teller said that he thinks the figure below has an area of 13.

a. What unit should be used to label the area of the figure? How do you know?

b. Explain how you might use a grid to check the area.

Student Mathematician: ______________________ Date: ______________

Measuring Using Cups

1. How many parts is the 1 cup divided into?

a. ______________________

This matches the ____________ color marks.

b. ______________________

This matches the ____________ color marks.

c. ______________________

This matches the ____________ color marks.

Student Mathematician: Date:

Measuring Using Cups

2. How much water is in each cup?

a. ______________________

b. ______________________

c. ______________________

d. ______________________

Student Mathematician: ______________________ Date: ______________

Measuring Using Cups

3. a. Shade in how $\frac{1}{2}$ cup of water looks.

b. Shade in how $\frac{3}{4}$ cup of water looks.

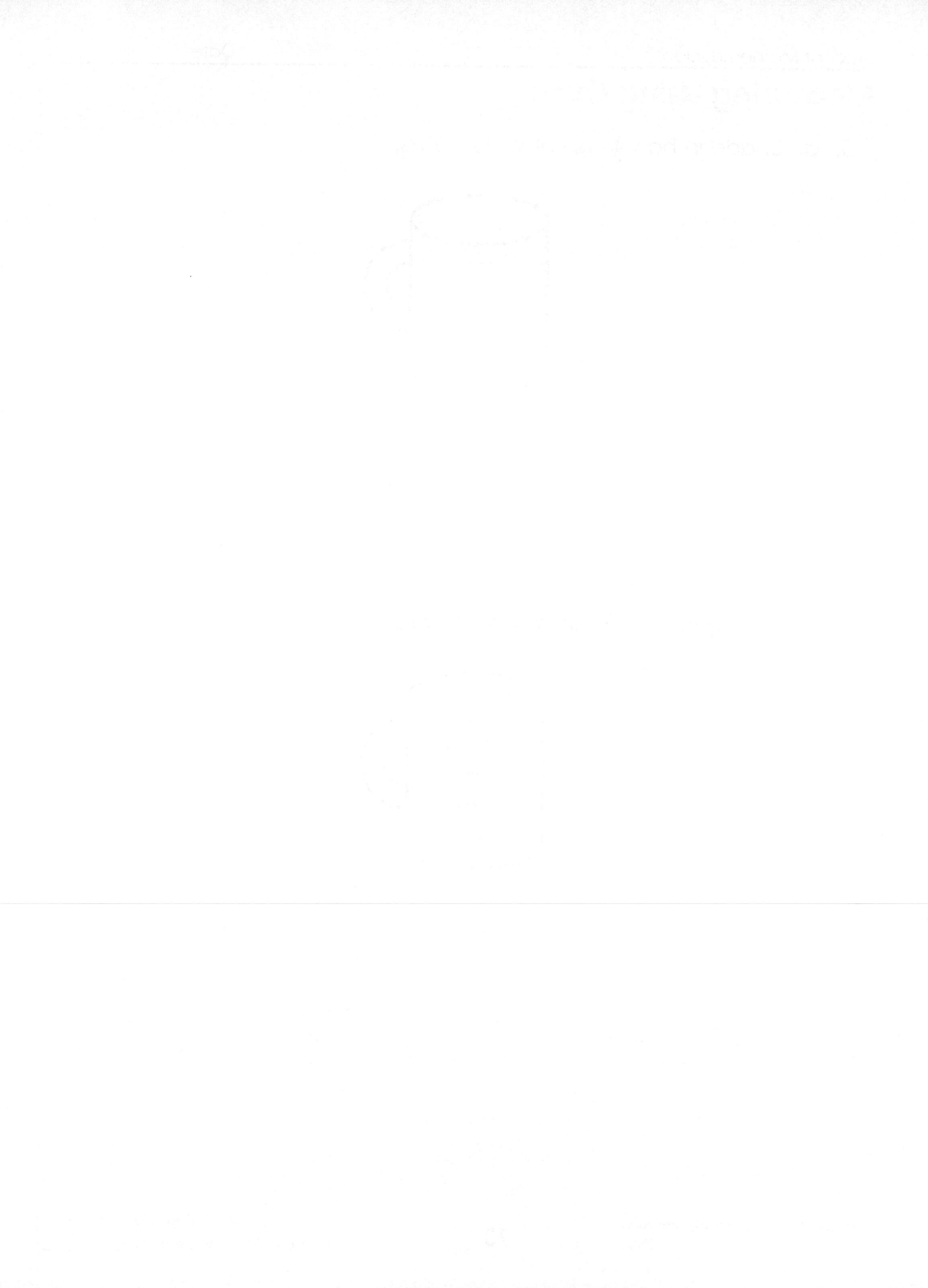

Student Mathematician: Date:

Measuring Capacity

1. Fill each container. Estimate the capacity. Then pour the material into the measuring container. Record the capacity of each container in the chart.

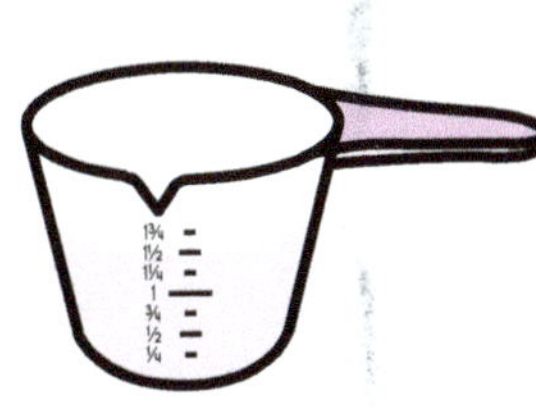

Container	Predict Capacity	Actual Measure
Example	$\frac{1}{2}$ cup	$\frac{3}{4}$ cup
A		
B		
C		

2. Which container holds the most? ______

3. Which container holds the least? ______

4. Put the containers in order from least to greatest capacity.

______ ______ ______

5. Find another container that holds less than any of the others.

What is its capacity? ______

6. Find another container that holds more than any of the others.

What is its capacity? ______

Student Mathematician: ______________________ Date: __________

THINK DEEPLY

a. Tell Dru and Teller how you measure the capacity of a container. Use pictures and words to explain.

b. Dru and Teller need to fill the container to 2 cups but have only a $\frac{1}{4}$-cup scoop. How many $\frac{1}{4}$-cup scoops will they have to use to fill the container to the 2-cup mark?

Number of $\frac{1}{4}$-cup scoops ______________

Student Mathematician: ______________________ Date: __________

Down the Drain Planning

Materials:

- ______________________
- ______________________
- ______________________
- ______________________
- ______________________
- ______________________

Plan:

1. ______________________
2. ______________________
3. ______________________
4. ______________________
5. ______________________

Draw your picture on the back of this page!

Student Mathematician: ____________________ Date: ____________

Down the Drain Data

Work Space

Water Left Running

Amount of water used to brush teeth:

____

Amount of water used in one day:

____

Amount of water used in three days:

____

Water Turned On and Off

Amount of water used to brush teeth:

____

Amount of water used in one day:

____

Amount of water used in three days:

____

Water Saved

One-day savings: ____________________

Three-day savings: ____________________

Student Mathematician: ______________________ Date: __________

Dru and Teller want to hear about the experiment so they can learn how to save more water.

a. Write a letter to them and explain how you conducted your experiment.

b. Explain to them how much water can be saved in three days if the water is turned on and off when we brush our teeth.

Student Mathematician's Glossary

Student Mathematician's Glossary

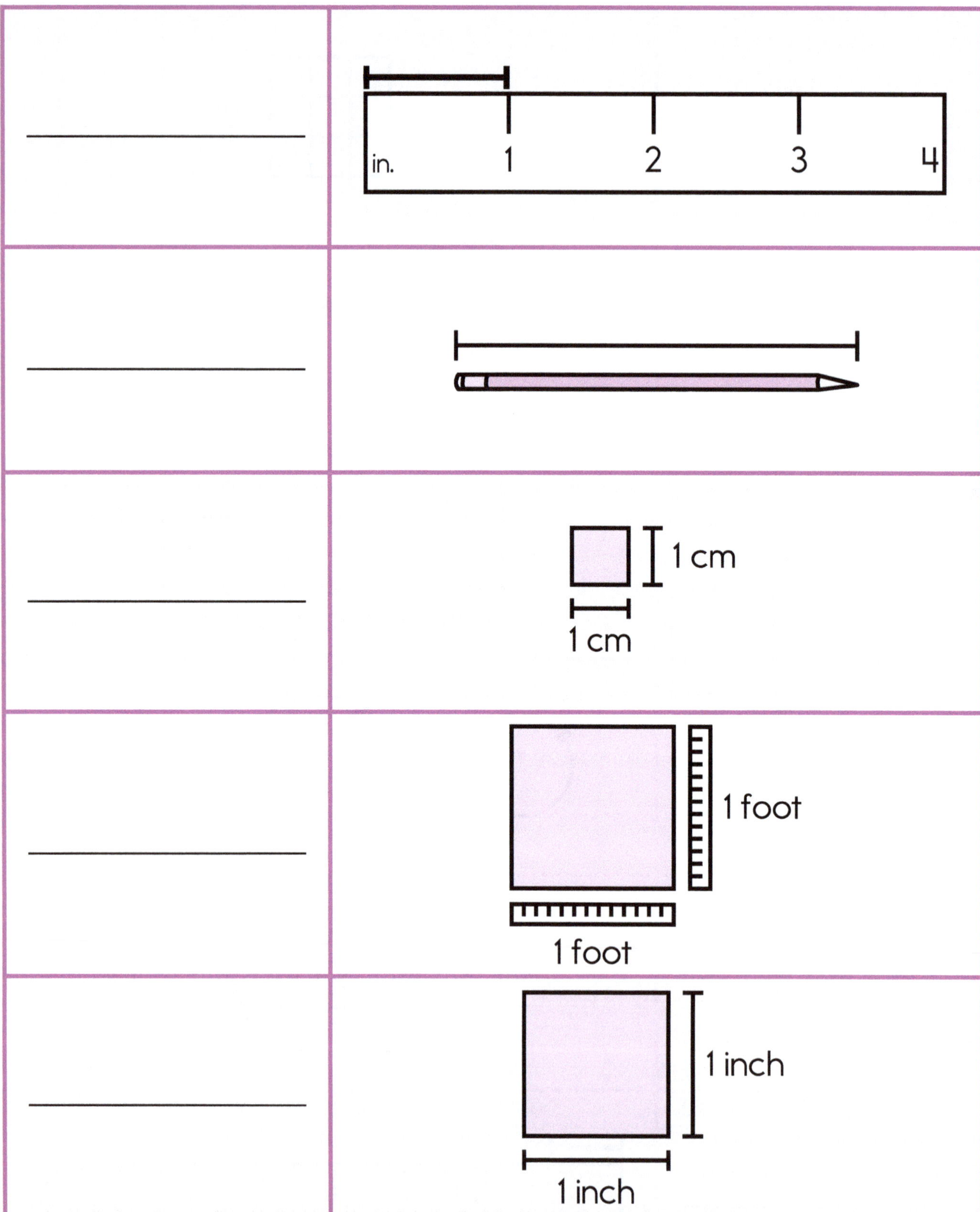

www.ingramcontent.com/pod-product-compliance
Ingram Content Group UK Ltd.
Pitfield, Milton Keynes, MK11 3LW, UK
UKHW051326070726
13610UKWH00014B/96